HEINKEL He 115

M2+HH of Kü.Fl.Gr. 106 at Weserflug in Einswarden

Torpedo/Reconnaissance/Mine Layer Seaplane of the Luftwaffe

H.P. Dabrowski

Schiffer Military/Aviation History
Atglen, PA

Translated from the German by Don Cox.

Copyright © 1994 by Schiffer Publishing, Ltd.

All rights reserved. No part of this work may be reproduced or used in any forms or by any means – graphic, electronic or mechanical, including photocopying or information storage and retrieval systems – without written permission from the copyright holder.

Printed in The United States of America
ISBN: 0-88740-667-X

This title was originally published under the title, *Heinkel He 115; Torpedoflugzeug - Aufklärer - Minenleger*, by Podzun-Pallas Verlag.

We are interested in hearing from authors with book ideas on related topics.

Photos:

B. Lange, F. Ritz, P. Petrick, Guhnefeldt via Petrick, H.J. Nowarra (deceased), K. Kössler, P. Ebel, V. Koos, F. Selinger, H.J. Meier, Kontakt, Heinkel Archives

Sources and Selected Reading

Caspari: E-Stelle See
Köhler: Ernst Heinkel, Pionier der Schnellflugzeuge
Kössler: Die He 115, das unbekannte Flugzeug (Modell-Magazin 3/78)
Green: The Warplanes of the Third Reich
Heumann: Heinkel He 115 (Flug-Revue)
Lange: Typenhandbuch der deutschen Luftfahrttechnik
Nowarra: Die deutsche Luftrüstung
Nowarra: Heinkel und seine Flugzeuge
Nowarra: Torpedo-Flugzeuge
Ritz: Mein Leben (unpublished)
Widfeldt: T2 Heinkel He 115 (Kontakt #77, June 1986)
Heinkel documents and He 115 archival compilations Heinkel factory newspaper from June 1939: "15 t Start nach Südamerika"
Vergleich He 115-Ha 140 (E-Stelle See Travemünde, 22 Sep 1937)
Entwicklung und Erprobung von Eisschwimmern für He 115 (E-Stelle See Travemünde, 30 Mar 1939)

M2+BL of 3/Kü.Fl.Gr. 106 with overpainted identification markings to cut down on reflection. The additional protective measure was often undertaken during minelaying operations.

Published by Schiffer Publishing Ltd.
77 Lower Valley Road
Atglen, PA 19310
Please write for a free catalog.
This book may be purchased from the publisher.
Please include $2.95 postage.
Try your bookstore first.

Front cover artwork by Steve Ferguson, Colorado Springs, CO.

NEMESIS OF NORWAY
The cover shows an He 115C of F.Gr.406 striking a convoy off the southern coast of Norway.

Introduction

The following information on the He 115 is based in part on a compilation of related material from the Heinkel Archives in Stuttgart. In addition, clarifications and highlights are provided to round out the picture of the He 115. At this point I would like to offer my heartfelt thanks to the Heinkel Company, Flugkapitän a.D. Dipl.-Ing. Friedrich Ritz and Dipl.-Ing. Karl Kössler, as well as all others who assisted me in this effort.

A New Seaplane

On 1 March 1935 the Luftwaffe was officially established as an independent branch of the German Wehrmacht. At that time the naval pilots had only the antiquated He 59 at their disposal for use as a twin-engine floatplane. A more modern airplane was needed, and based on a 1935 requirement issued by the Reichsluftfahrtministerium (RLM), the Heinkel Flugzeugwerke produced a twin engine multi-purpose seaplane towards the end of 1936 as a replacement for the He 59. Hamburger Flugzeugbau developed the Ha 140 as a competing design. Heinkel's product was given the designation He 115. Siegfried Günter, the designer, developed the He 115 as an all-metal mid-wing aircraft, which initially was powered by two BMW 132 N engines producing 865 hp each (later, as with all follow-on models, the aircraft was equipped with BMW 132 K engines at 960 hp each). The operational use for the aircraft: long-range reconnaissance, torpedo bomber, minelayer and fog dispenser.

The structural sturdiness was evaluated in accordance with Beanspruchungsgruppe H 3, during which particular attention was given to its seaworthiness. Testing of individual components took place in the spring of 1937 using an He 59 D modified by Arado. This aircraft flew with a canopy mock-up for determining the most favorable configuration for the machine gun position in the fuselage nose. A cage-like design resulted, which had a traversing cupola attached for the machine gun. This design reflected the commonly held concept which called for the gunner to man the machine gun in either a sitting or half upright position. A specific height was inherent in such a design, however, and the complete subassembly didn't entirely fit in with the otherwise elegant lines of the He 115.

The maiden flight of He 115 V1 (D-AEHF) took place in August of 1937. Initially, Heinkel chief test pilot Gerhard Nitschke was responsible for the aircraft, but after a only a few flights Dipl.-Ing. Friedrich Ritz took over, flying the majority of production aircraft models as well. Factory test pilot Ritz had these comments on the He 115:

"The bird was cursed with absolutely terrible flight characteristics, which so horrified a pilot of the Ministerium (given an early unofficial flight by Heinkel) that he sent a castigating report to the Technisches Amt. This in turn caused Udet to write Heinkel that the He 115 would never fly with the Luftwaffe under such circumstances.

With this as an impetus, there was serious work undertaken to make improvements. Success was finally achieved by using so-called "power-controlled auxiliary control surfaces", which were control tabs requiring less physical strength to operate and which in turn brought the main control surfaces to the correct position. The same pilot, which had been so disgusted before, now filed such a favorable report after a follow-on visit four weeks later that Udet was full of praise and

In front is the predecessor, the He 59, while in the background is its successor, the He 115.

The He 115 V1 is ready; aside from the "HEINKEL" name, it carried no other markings.

the He 115, which would later go into production, became extremely popular with the naval flyers."

The He 115 V1 and V2 (D-APDS, first flight in November 1937) were initially fitted with the ribbed nose glazing and entered testing with such. However, at the same time, the design bureau was already working on an improved, sleeker, 'horizontal' nose canopy (where the gunner would lie prone), a feature which was introduced from the third V model onward.

First Comparison

As early as 22 October 1937 a theoretical comparison had been made between the He 115 and the Ha 140 by the Erprobungsstelle (or E-Stelle) See in Travemünde. By examining the various characteristics of the two designs, it was determined that they were roughly on par with each other.

Without going into an in-depth discussion of the Ha 140, there should nevertheless be mentioned the fact that its first flight took place on 30 September 1937, approximately one month after the He 115 V1. Shortly thereafter, it was lost following a hard landing. The V2 was finished in November 1937 and was the aircraft which underwent practical testing in the Lübeck Bay in February of 1938. Helmut Wasa Rodig was at the controls of the Ha 140, while Friedrich Ritz flew the He 115.

Friedrich Ritz gives the following account of how the comparison progressed: "It was planned that we would make the required three landings one after the other, Wasa always going first."

"I was therefore able to observe his first landing in the Travemünder Bay, on the turbulent Baltic waters, as well as his subsequent takeoff. I noticed that the takeoff run was extremely long and that the Ha 140 only became airborne after overcoming 8-10 swells, meaning that its takeoff speed was significantly greater than that of the He 115, which was in the air after 3 swells!"

"I then completed my first landing and takeoff without any problems, then prepared myself to watch Wasa's second landing. I was hardly surprised when the Ha 140 was damaged in this landing."

"The tests were broken off at that point, Wasa was picked up by the rescue ship, and I landed without incident on the Pötenitzer Wiek."

"On one side of my aircraft a tension wire had torn free, an occurrence which I initially kept to myself; the tension wires, which on the V1 linked the float supports with the fuselage and wings, were quietly replaced by stronger support braces."

Eight World Records

The test flights with the new seaplane then proceeded smoothly, for there were plans to use the design – with minor modifications – in establishing record flights. Accordingly, the He 115 V1 was given a sleek new wooden nose profile and a slender canopy which was blended into the upper fuselage.

On 20 March 1938 He 115 V1/U1 took off in Warnemünde and, after nearly an hour's flight with a 2000 kg load, passed over the starting point at an altitude of roughly 4500 meters. The route for the record-making flight was via three points: Laboe (near Kiel) –Swinemünde – Leba (Pomerania). With this flight (back and forth over the route twice), lasting a total of 7 hours 15 minutes, eight records would be established in one fell swoop. The record flight was carried out by Dipl. Ing. Friedrich Ritz, accompanied by the mechanic Schmid from BMW. The following data shows the speeds at which the new records were set:

World Record for Speed with:

0 kg	load over	1000 km –	330.615	km/h
500 kg	load over	1000 km –	330.615	km/h
1000 kg	load over	1000 km –	330.615	km/h
2000 kg	load over	1000 km –	330.615	km/h
0 kg	load over	2000 km –	328.467	km/h
500 kg	load over	2000 km –	328.467	km/h
1000 kg	load over	2000 km –	328.467	km/h
2000 kg	load over	2000 km –	328.467	km/h

Just 11 days later these records were broken by an Italian CANT Z 509, but this aircraft was powered by three engines totaling 3000 hp takeoff performance, compared with the He 115 with only half the power.

Nevertheless, the record-setting flight of the He 115 V1 clearly showed that the improved canopy significantly increased flight performance.

Failed Record Attempt

In March of 1939 an attempt was made to risk the record-breaking airplane in a hop across the Atlantic to South America. Accordingly, the covered observer's position in the rear of the fuselage was opened and equipped with auxiliary controls.

The gross weight was increased by nearly 6 tons to 15 tons due to the additional fuel required. The crew was comprised exclusively of experienced personnel "loaned" from Lufthansa: Flight Captain Walter Diele, Radioman Paul Geisler and Engineer Simon Butz.

The heavily laden He 115 V1 took off on its South American journey on 14 March 1939 at 14:06 GMT from Riebnitz with the help of a "tow plane" in front. Unlike previous Atlantic crossings, this one was to be a non-stop flight.

The plane, which was scheduled to cover a total distance of over 10,000 km, passed over Heligoland, the English Channel and the Bay of Biscay, then continued via the Canary Islands. The altitude during the day was 20 meters, and at 16:50 hrs GMT the following day the plane passed the German steamer "Monte Pascoal" at some distance. A short time later the pilot, Walter Diele, noticed a wide streak of oil coming from the port engine and spreading out over the wing. Diele turned the plane around, back toward the "Monte Pascoal." The catapult ships "Ostmark" (Bathurst) and "Friesenland" (Pernambuco) were notified, the German main control center was informed and given regular updates on the situation.

Bad luck! With a collapsed float and its starboard engine broken from its mounts, the Blohm und Voss Ha 140 lost any chance for production orders.

The He 115 V1 before and after modification – the differences are clearly discernible here: new nose profile, new canopy, altered rudder and the absence of the boarding ladders leading up to the fuselage.

Right: The record-setting bird on the cover of the Heinkel-Werkzeitung from February/March 1939. As opposed to other aircraft, such as the He 100, as a record-breaker the He 115 was not used as a means to deceive the public or held back for propaganda purposes.

Below: The He 115 V1 over the Mecklenburg landscape following its modifications to a record-breaking airplane.

The aircraft was boarded by means of a collapsible ladder since the wire ladders were removed for aerodynamic reasons. Unlike the Ritz version, the navigation compartment at the rear of the center fuselage was equipped with auxiliary controls "in case of need."

Pilot Walter Diele readies the He 115 V1 for its Atlantic crossing.

Below: Takeoff on its journey across the Atlantic. The overloaded V1 needs a "towplane" to get airborne, a procedure which was practiced several times before the actual flight. It is unknown why the front aircraft is painted with a light colored fuselage and dark wings.

End of the attempt on the record: Walter Diele was forced to set the He 115 V1 down near the German steamer "Monte Pascoal" and continue the remainder of the journey with his crew on board the ship.

He 115 V2, the last aircraft to be fitted with the older style A-Stand, or forward machine gun position.

From the front the only discernible difference between the V2 and the first V1 is the former's smooth engine housing.

The V2 seen during ice runner testing in March 1939 in Sweden. This testing was conducted by members of E-Stelle See, Travemünde.

Butz determined that the port engine would only run for a few more minutes, which prompted Diele to climb to 350 meters and continue flying with only his starboard engine. Several thousand liters of fuel were dumped, the throttled engine shuddered a few times and then the propeller came to a stop. It wasn't possible to continue under such conditions, the "Monte Pascoal" was notified, and D-AEHF landed near the steamer at 17:36 hrs GMT in seas of 4-5 after completing 2/3 of the total flight, or 6700 km. A broken oil line had ended this flight.

Aircraft and crew were lifted on board, and eventually the goal of Rio de Janeiro was reached – even if it wasn't in the planned manner. Although the flight wasn't successful in terms of establishing a new record, the performance of the aircraft was truly remarkable.

Let's now take a look at the other V-series: the He 115 V2 was still equipped with the older ribbed nose glazing and was therefore used for testing sighting mechanisms; aside from reconnaissance duties the He 115 was also planned for use as a torpedo bomber. In addition to the normal Lotfe-Gerät the V2 also was fitted with specialized equipment for dropping torpedoes.

In the time frame between 6 March and 21 March 1939 this aircraft was also used to test reinforced floats for operating in ice and snow. The He 115 B-2 series was later equipped with these ice floats, which will be discussed later. This testing was conducted by E-stelle Travemünde in Sweden. For this purpose steel runners were installed on the undersides of the floats on the V2, in order to facilitate landing and takeoff on ice and snow. A production aircraft was not available at the time, so in view of deadline requirements the 2000 kg lighter V2 was utilized (although it still had the wire-braced float arrangement).

He 115 V3, D-ABZV, the first model with the newer glazing, which basically remained unchanged with all further He 115 designs.

Below: He 115 V4, D-AHME, seen with its enlarged rudder. It was the last design with the tension wire-braced floats.

The He 115 As diverted for "Mediterranean Corrosion Testing" – the only two airplanes to serve in the Spanish Civil War.

26 March 1939: an He 115 A in Puerto di Pollensa, Mallorca.

Below: Engine maintenance after the arrival of the "Spanish" He 115 As.

The fuselage nose compartment was a series-produced component, seen here in two stages of completion.

 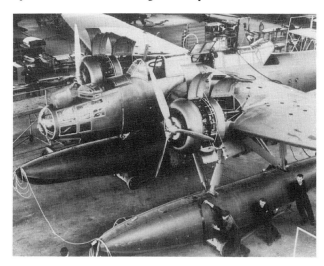

The He 115 in production, in a nearly completed stage.

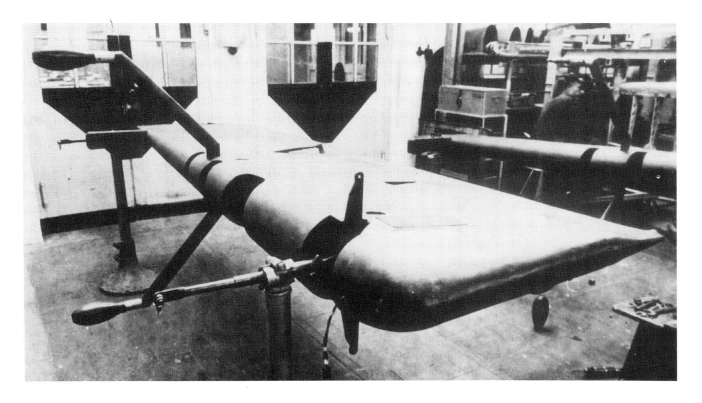

Manufacture of the vertical stabilizer, which were fitted with the prominent counterbalances beginning with the B-series.

A BMW 132 K engine suspended from a crane for the He 115 production line. At Heinkel the He 115 was manufactured in the same assembly hall as the He 111.

HE+PA in flight, the letters HE behind the Balkankreuz indicating a factory-operated Heinkel plane.

A Swedish military delegation inspecting one of the planes ordered by their government prior to acceptance.

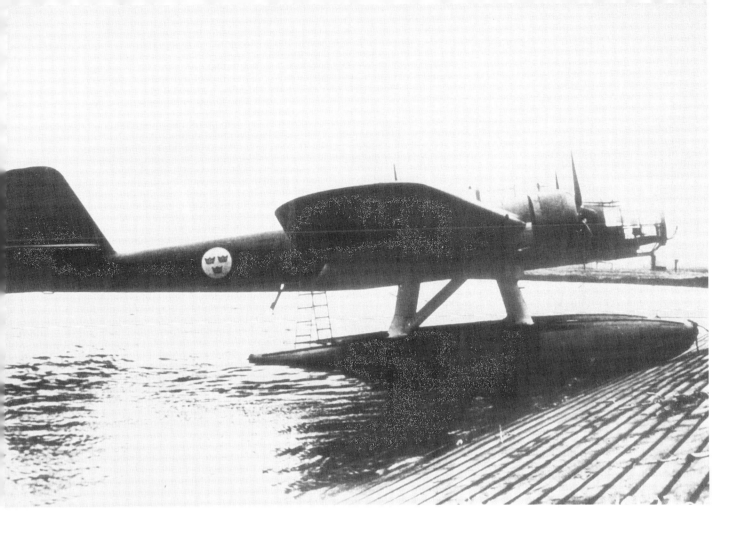

A Swedish He 115 on the launch slip and during takeoff.

The first He 115 delivered to Sweden crashed and burned in Hägernäs on 13 August 1939 while demonstrating a turn powered by only a single engine. The three crew members were killed in the crash. (Kontakt)

A Swedish He 115 releasing a torpedo in 1944. (Kontakt)

From Rostock-Marienehe the aircraft was flown to Stettin (interim landing), then to Kalmar (Sweden), continuing to Östersund and Frösön, with another interim landing in Stockholm Hägernäs. Near Frösön thorough testing was conducted using the ice floats. After 25 ice landings it was back to Stockholm on the 18th of March 1939. There the aircraft was test flown by representatives from the Air and Naval Ministries. During the stopover in Stockholm the He 115 was tethered to the buoy a total of 42 hours and, although a Swedish pilot accidentally landed on a 3 cm thick section of drift ice (which echoed loudly) during the testing, there was no major damage and the floats held fast. Even though not all the testing conditions were available (for example, there was no bare ice), the verdict was a favorable one and the ice floats would, with minor modifications, be used on production aircraft.

The following test aircraft, the He 115 V3 (D-ABZV), was fitted with the newly developed nose glazing, which would also be used on subsequent series-produced aircraft. The He 115 V4 (D-AHME) was initially given a new rudder, as found on the A-series, and later equipped with a new float support arrangement. The floats were no longer supported by wire bands, but were held to the fuselage and wings by two shrouded braces per float, thereby dispensing with the necessity for tension wires running to the outer wing panels. With this arrangement, the V4 served as the prototype for the first 0-series production models.

The A-0 series totaled 10 aircraft, which were delivered to the Luftwaffe starting in January 1938, where they initially found use as reconnaissance platforms. The general design of the A-0 series carried over into the following series and remained in production until 1944. The fuselage was of a tapered design with an oval cross-section.

The Swedish He 115s were fitted with the Ksp m/22 – 37 R, 8 mm machine gun, since Heinkel was not permitted to supply the aircraft to foreign countries with armament. (Kontakt)

Norwegian He 115 minus armament.

Above: Ernst Heinkel poses for the cameraman with the Norwegian delegation in front of one of the last two He 115s received by this nation.

Left: a few interesting details can be seen here: far left at the base of the float support is the opening for the fuel dump, the wire ladder with metal supports and the sprung steps on the top of the wing.

The last Norwegian He 115s leave the Heinkel harbor.

He 115 M2+EH of 1/Kü.Fl.Gr. 106, in which Lt. z. See Joachim Vogler was forced to make an emergency landing on 13 April 1940 on Lake Sola near Bodó due to lack of fuel. It was captured by the Norwegians and given the designation of F.62.

He 115 A-2 (Norwegian F.50) as a Finnish seaplane (Kontakt)

The same aircraft seen from a different angle.

Above: the former Norwegian F.56, now BV 186 in service with the Royal Air Force. (Kontakt)

An He 115, operated by the RAF, is removed from service. In December 1942 this aircraft met its end by scrapping in Woodhaven, Scotland, south of Dundee.

The end of the RAF He 115: breakdown into large component parts and finally chopped up for the smelting oven.

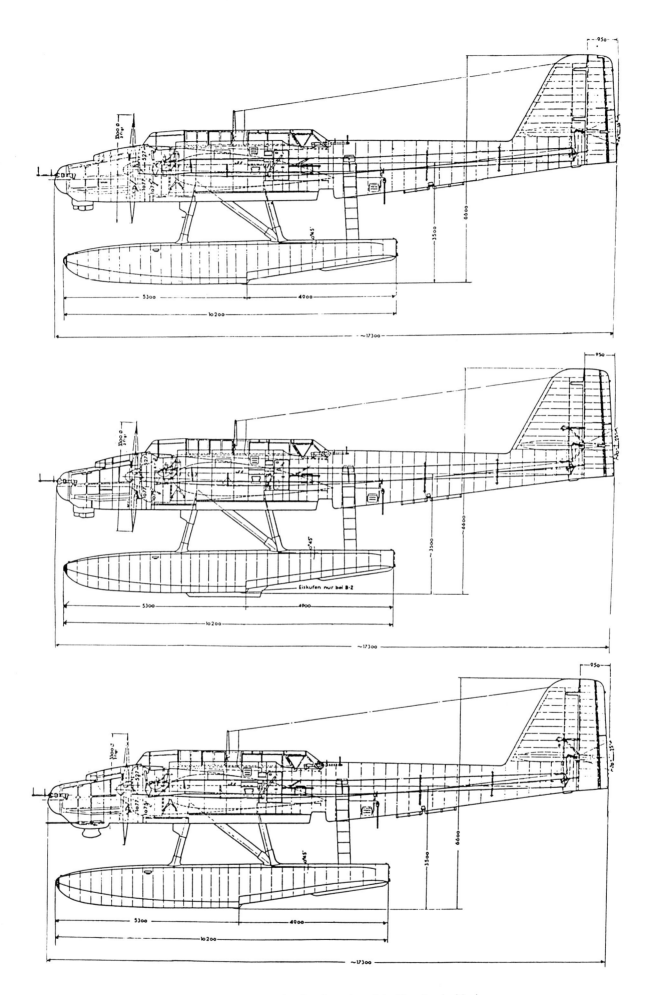

The basic design of the He 115 A, B, and C-series (top to bottom), taken from the original handbook sideviews.

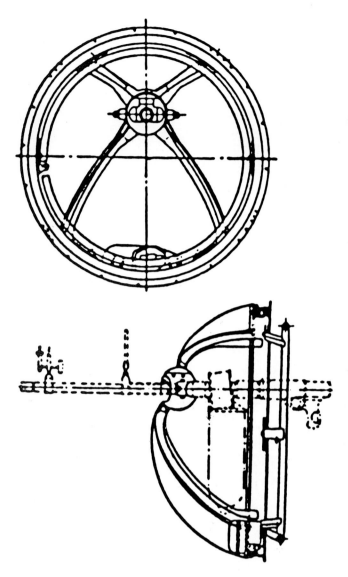

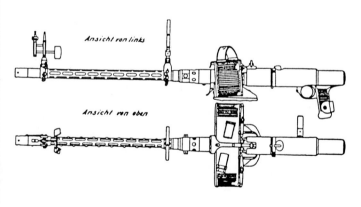

The Ikaria GD-A 1114 cupola (above left); the field of fire for the machine gun was 120 degrees. The photos show the crew station in the nose of the He 115, both occupied and unoccupied. The MG 15 was the standard gun for the A-Stand's GD-A 1114 cupola.

The following Heinkel factory pilots have gathered beneath the wing of a production He 115 (left to right): Kurt Heinrich, Friedrich Ritz, Hans Dieterle, Gerhard Nitschke and Fritjof Ursinus.

Friedrich Ritz during a test flight in a production He 115. The seat could be lowered approximately 1.5 meters straight down to facilitate better visibility during torpedo attacks. During one test flight this seat inadvertently dropped "into the basement", a situation which nevertheless didn't interfere with the landing approach.

A rather large group of He 115s, which was an uncommon sight.

The He 115 in the background is TI+HD, former V5 D-ABBI, photographed at E-Stelle See, Travemünde.

The Heinkel launching team manhandles the DC+GI and DC+GH.

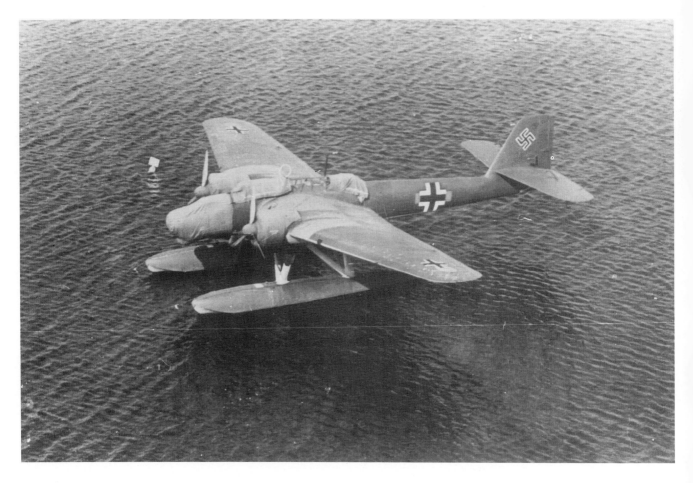

Here all weather sensitive parts are carefully covered with canvas tarp.

It was of monocoque construction, with open formers and main bulkheads, and was given bolt connections for the specially produced 'horizontal' nose glazing. The extended canopy for the pilot and radio operator was set on the fuselage above the wings and offered adequate visibility on all sides thanks to its extensive glazing. The pilot's seat could be lowered approximately 1.5 meters to provide better visibility forward during torpedo runs. The weapons bay was located on the fuselage underside beneath the wings. It could carry droppable ordnance in the form of torpedoes or a correspondingly heavy bomb load. Later torpedo bomber models were sometimes fitted with external racks for carrying two torpedoes; these racks could also be used for mines as well.

The twin-beam cantilever wing consisted of a center section and the two removable outer pieces. The rudder was also cantilever, but the horizontal stabilizers were fastened to the bottom spar of the fuselage with a cross-brace. The auxiliary control surfaces, or tabs, could be adjusted from the pilot's seat during flight. The flotation system consisted of two metal alloy floats, which, as already mentioned, were supported by braces. The engine for all versions was the BMW 132 K, delivering a takeoff performance of 960 hp and an operational performance of 830 hp.

While deliveries were still underway with A-0 series aircraft, manufacture had already begun on the first of the A-1 models, of which a total of 38 would be built. These were all delivered to the Luftwaffe in 1938. This model dispensed with the racks (PVC 1600) for the torpedoes, as it was planned to utilize this model as a bomber, fog dispenser and long-range reconnaissance aircraft. Compared with the A-0 version, the A-1 possessed a higher takeoff weight of a maximum of 9600 kg (9100 kg for the A-0). The weapons bay contained racks for 3 250 kg bombs, while additional 250 kg bombs could be carried on two hardpoints beneath the wings. This version was given the designation He 115 A-1/R1.

Foreign Orders

An export series was produced from the A-1 production run, given the designation A-2. These aircraft, however, were fitted with a tail assembly more akin to that of the B-series (with counterbalances). During the time period between 20 June 1939 and 20 October 1939 Sweden received 12 of a total of 18 He 115 A-2s. The remaining 6 planes were not delivered due to the war's interference. These aircraft were designated in the Swedish Air Force as T2 (T = torpedo bomber). The first of the delivered aircraft crashed after its 80th flight hour on the 13th of August 1939 during a flight demonstration in Stockholm-Hängernäs. The plane was consumed by fire and the crew perished in the crash. The number of Swedish He 115s gradually diminished, so that by 1949 the remaining four aircraft were written off due to the high operating expenses; the last example was withdrawn from service on 13 December 1952 (!) after accumulating 1702 hours flight time.

Norway ordered 12 aircraft of the A-2 production batch, with 6 of them being delivered by the end of 1939/beginning of 1940. Starting on 9 April 1940 Norway became an enemy nation, precluding delivery of the remaining He 115 planes.

The Norwegians numbered their aircraft in increments of two: F.50, F.52, F.54, F.56, F.58, and F.60 were purchased from Heinkel, while F.62 and F.64 later fell into Norwegian hands. The latter two aircraft were B-1s from 1/Kü.Fl.Gr. 106. F.62 was stationed in Leirvika near Hammerfest beginning on 6 May 1940 and was piloted by Senior Lt. Carl August Stansberg with Senior Lt. Alf Steffen-Olsen acting as observer. An American volunteer for Finland, Donald K. Willis, often flew along as gunner.

F.62 flew its last mission on 20 May 1940, then continued to Skattöra near Tromsö for overhaul, where it was recap-

He 115 C with the latest coat-of-arms for 2/Kü.Fl.Gr. 506, armed with an MG 151/15. The belt feed for the forward-firing fixed gun is clearly visible.

Gerhard "Francois" Nitschke, Heinkel's chief test pilot, is seen here climbing into a production He 115 in his unmistakable leather jacket.

Maintenance work on He 115 CA+BW.

tured by the Germans when they took the town on 13 June 1940. F.64 was able to escape to England in time before the Germans could retake it. F.52, F.54 and F.56 also set out for England; F.54 wasn't fortunate enough to complete the voyage and was forced to make an emergency landing off the Shetland Islands.

The aircraft, now designated BV 184 through BV 187, were fitted with auxiliary fuel tanks, more powerful guns and a modified canopy. The armament was increased to four rear-firing guns and four forward-firing guns, along with fixed machine guns mounted in the wings; a section of the canopy was covered over. Two of these modified aircraft (BV 184 and BV 185) operated out of Malta with Norwegian crews, sometimes with German markings for use on secret missions. The two planes were sunk during a bombing attack on Malta. BV 186 and BV 187 flew similar missions against Norway from Scotland. At the end of 1942 they were scrapped – in Woodhaven, Scotland, from where they had conducted their operations.

F.50 flew to Finland, which only became Germany's "comrade in arms" beginning on 26 June 1941. F.58 escaped to northern Norway, where it continued to fight against the Germans. F.60 fell into German hands near Stavanger.

In September 1938 the main production run for the A-3 series began, a version which was to be delivered exclusively to the Luftwaffe. The operational use of this model was the same as for the A-1 series; however, the weapons bay was configured to accept an additional droppable tank for 50 liters of fuel. Otherwise, the design was the same as that of the A-1.

The He-115 in the Spanish Civil War

Near the end of the Spanish Civil War two He 115s (D-ANPT and D-AHOS) were flown to the Spanish Mediterranean island of Mallorca. Purpose: "Mediterranean Corrosion Testing." Responsible for the undertaking was Fliegerstabsingenieur Dipl.-Ing Gerhard Geike ("Teniente Colonel" during the Spanish deployment) from the E-stelle See Travemünde; he also flew one of the two He 115s to Mallorca.

Photograph of a torpedo on a raft. Since the He 115 was often moored far out in the water prior to a mission, this procedure was a common occurrence. Refueling was also often done in the same manner.

A machine is made ready for takeoff and the canvas tarps are removed. It is not clear what purpose the commonly seen cable running along the underside serves.

Right: An He 115 C of 3/Kü.Fl.Gr. 106

Rescue practice with PP+AX in Pütnitz.

S4+AK of 2/Kü.Fl.Gr. 506 with all lightly colored surfaces painted out.

Left: M2+BL also has all its lighter colors painted over. This aircraft is equipped with rear-firing MG 17 guns behind the engine housings.

Here an He 115 in the winter Norwegian theater is protected against prying enemy eyes with white painted areas.

An He 115 at the Heinkel factory.

Takeoff of an He 115; notice the fuselage Balkankreuz with its narrow outline of white.

CA+BV in flight.

He 115 C VF+UY in flight.

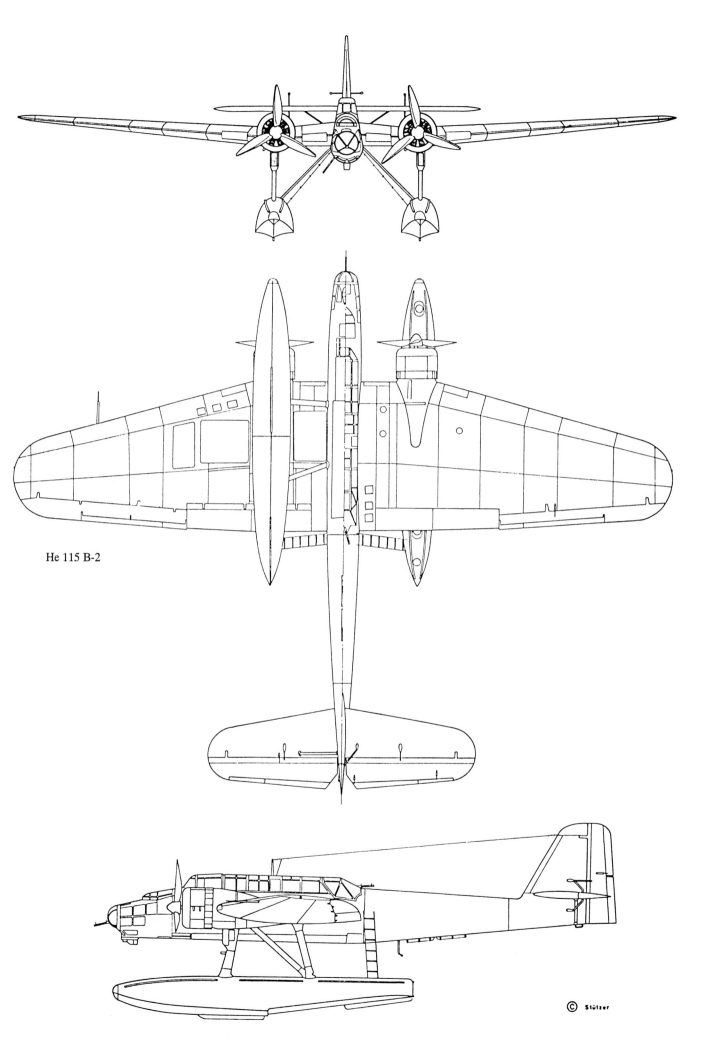
He 115 B-2

He 115 B-2 CA+BR takes off from a frozen lake with the aid of reinforced ice runners beneath the floats.

Left: CA+DU, also a B-2 version, takes off in 1942 from the ice-covered surface of the Zwischenahner See.

Additional participants were: on-board mechanics Burmeister and Dortmund, radioman Schröder (all from E-stelle), Lt. Fiehn, Lt. Jandray, Lt. Wachsmuth, plus additional crew members and ground crew, all from the Luftwaffe.

With Luftwaffe crews, the planes set out at the beginning of 1939 from Rostock-Marienehe to Travemünde, given last-minute checks, then continued (in civil markings) via Friedrichshafen and Italy, finally arriving at Puerto di Pollensa on Mallorca where Aufklärungsstaffel 88 of the Legion Condor was stationed. In gray paint and with Spanish markings the two He 115s were flown on several reconnaissance missions without enemy contact. During a demonstration before Spanish officers in April of 1939 (the exact date is unknown, since no flight logs were kept in Spain) Dipl-Ing. Geike performed a loop with one of the two aircraft in Puerto di Pollensa.

A few Heinkel representatives were at this demonstration, which led to intents to purchase. The two He 115 As returned back to Travemünde in May of 1939, flown by Luftwaffe crews. The deployment in Spain was so short that the aircraft were not even painted with the customary Spanish designations. Final investigation determined that the two airplanes had suffered an inordinately high amount of corrosion damage during their deployment.

Various Roles

The He 115 V5 (D-ABBI) can be considered the prototype of the B-series. Following a short run of B-0 models, at the end of 1939 the A-series was replaced by the B-1 series, which was specifically equipped for operations as a torpedo bomber and for dropping mines. By strengthening the design the take-off weight was brought up to a maximum of 10815 kg. In its role as an aerial minelayer the aircraft received dropping equipment for releasing LMA III and LMB III mines. Four modification versions stemmed from the basic B-1 series: He 115 B-1/R1 as a reconnaissance platform without the Lofte-Gerät and with cameras installed in the weapons bay, He 115 B-1/R2 as a bomber with 1 x SC 500 or SD 500, He 115 B-1/R3 as an aerial minelayer, and finally as a fog dispenser with an SV 300 device. The range was 2600 km, with the reconnaissance version achieving 3100 km due to its reduced weight.

The B-1 series spawned the B-2, which was identical in every respect except for the addition of ice runners underneath the floats. For testing purposes, one He 115 had its entire flotation system replaced by snow skis, but the ice runners proved

He 115 M2+HH of 1/Kü.Fl.Gr. 106 hung from a crane at Weserflug in Einswarden on 3 August 1940. WFG did not license build any He 115 aircraft, but instead played a major role in making modifications based on mission requirements.

An He 115 B of the Fliegerwaffenschule (See) 2 in Bug on the island of Rügen, 1939. Notice the high-visibility paint scheme on the floats.

Right: An MG 151/20 20 caliber machine gun fixed in the nose of an He 115.

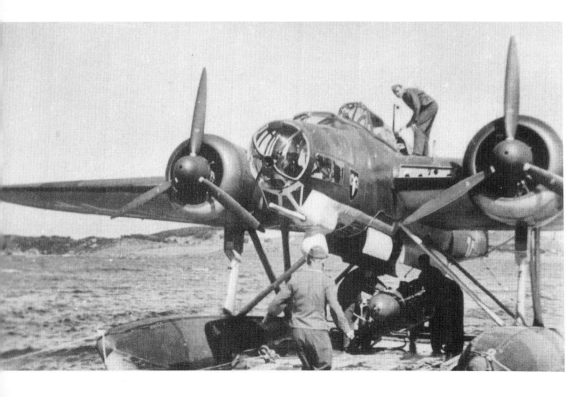

An He 115 C-1 of 3/Kü.Fl.Gr. 506 is loaded with a live torpedo.

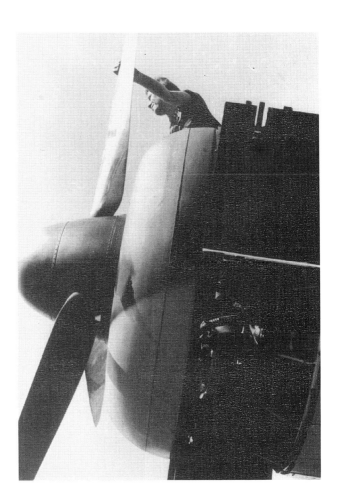

The BMW 132 K engines of the HE 115: fuel injected, takeoff performance rated at 960 hp, 9 cylinders. The two intake valves on the left and right sides of the uppermost cylinder were designed to make use of the compressed air to enhance performance. In its version as a single point fuel injection design, the BMW 132 K was only delivered to Luftwaffe aircraft. The propeller diameter of the VDM propeller was 3.3 meters.

more successful since they could operate from frozen water. Two special prototypes were fitted with airbrakes for shortening the landing roll; these were similar to dive brakes which extended above and below the wings. For the same purpose a single He 115 was equipped with fully adjustable propellers, a solution which offered the best results.

In a message dated 18 January 1940 from the RLM to the Heinkel Werke, mention was made of an He 115 V8 with an enlarged weapons dropping mechanism – this would have to be the last of the V prototypes, for development of the D and E series was brought to a stop and the He 115 C model production run was allowed to run out. Between the V5 and the V8 there must have been a V6 and V7, but nothing is known of these aircraft. During the same year the B-series gave way to the more heavily armed C-series. Two subelements of the C-series existed: the He 115 C-1/R1 as a bomber and the He 115 C-1/R-2 as a reconnaissance model. The A-Stand, or forward gunner's position, received an additional MG 151/20 with 200 rounds of ammunition, and the Lofte-Gerät was given a teardrop-shaped shroud.

The C models could be fitted with two rear-firing MG 17 machine guns in the engine nacelles, as could models from the B-1 on (with a field modification kit) beginning on 1 September 1942. In its role as a torpedo bomber, two LF-5W torpedoes could be carried beneath the fuselage. With this arrangement, this version had the greatest all-up weight of all, with 11100 kg. The He 115 C-2 was identical to the C-1, but as with the B-2 it was also equipped with ice runners. This model was accepted by the Luftwaffe beginning in 1940. The next version in the C-series was the He 115 C-3, a special model for operations as a minelayer.

Engine maintenance, one of the most important tasks. The gauges for rpm, pressure, temperature, etc. could be read directly by the pilot – via a viewing glass on the engine housing.

In the rough conditions of the polar theater it was quite easy to damage a float. The torpedo racks of the He 115 could be used to quickly transport a replacement float. However, the aircraft became extremely clumsy in such a configuration, which meant that this arrangement could only be used in extenuating circumstances.

An He 115 in Norway, with larger type German ships providing the background setting.

This He 115 used in Norway was also equipped with the rear-firing guns.

The armament was the same as the C-2, but the B-Stand (the rear gunner's position) was given additional armor plating. 18 of this type were used operationally, which in 1940 and 1941 successfully dropped mines around Great Britain. The missions were carried out at night, so that for a long time afterward these activities went unknown. The last version of the C-series was a torpedo bomber model, which entered service as the C-4 beginning in May of 1941. A total of 30 aircraft of the C-4 version were constructed. These aircraft only carried a single gun in the A-Stand, either a traversible MG 15 or the fixed MG 151/20. A single aircraft from this series was fitted with twin BMW 801 engines with the purpose of operating as a high-speed long-range reconnaissance aircraft. This airplane, designated the He 115 D, was operational in 1941 and was given the code PP+ND.

It was originally planned to power the He 115 D with two BMW 800 engines. However, since this engine never went into production, it was decided to use the significantly more powerful BMW 801 engine. This engine was needed exclusively for fighter manufacture (Focke-Wulf Fw 190) and was therefore ruled out for the He 115 D-series production. The last series model was the He 115 E, which entered production in 1942 – a few examples were manufactured until 1944, which is rather remarkable since development and construction was officially halted (along with that of the He 115 D) on 18 January 1940. Apparently such a shortage of this type was felt that it became necessary to reopen the production lines, most certainly as a result of the constantly changing priority requirements. The He 115 E possessed an improved bombsight, while a few of these flew unarmed as the He 115 E-2 in the air-sea rescue role.

Which units flew the He 115? The Küstenfliegergruppen 106, 406, 506, 706, and 906 made use of this plane, but it was also utilized by training units (such as Fliegerwaffenschule (See) 2 in Bug/Rügen), with KG 200 or even as a tow plane for a float-equipped DFS 230 glider tested on the Chiemsee in Bavaria! In the fall of 1944 this pair was caught by low-flying enemy fighters and shot to pieces.

The numerous missions flown by the Küstenfliegergruppen would certainly necessitate a lengthy chapter of themselves and cannot be presented in a detailed format here. The He 115s participated in the attacks on the Western Allies' PQ convoys attempting to supply the Soviets – with mixed success. There were also a series of spectacular individual cases. A short selection: flight with landing and return to Jan Mayen (12 Sep 1940), then departing for final time; collapse (17 Sep 1940); a flight to Narvik (April 1940), where the He 115 was hidden for four weeks in the last fjord still in German hands, until fuel for the return flight could be arranged; rescue of five Englishmen far out in the Atlantic (23 June 1941).

A total of approximately 400 He 115s were built up until 1944. At the war's end the Allies found seven examples in Germany and one in Norway. These were destroyed; it is not known whether a complete aircraft of this type survives today anywhere in the world.

Postcard perfect and reality: successful torpedo runs as portrayed here were rare. On the other hand, friendly losses at the hands of enemy aircraft could be expected at any time – as here in the area of Bergen on 6 February 1944.

Also from the publisher

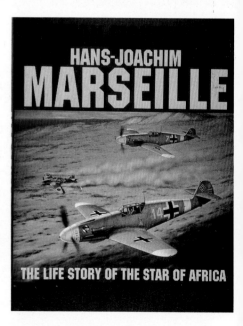

Please write for our free catalog which includes over 200 military and aviation titles.
Schiffer Publishing Ltd., 77 Lower Valley Road, Atglen, PA 19310